INVESTIGATE EARTH SCIENCE

Barbara Allman

LAKES and RIVERS

Enslow Publishing
101 W. 23rd Street
Suite 240
New York, NY 10011
USA
enslow.com

Words to Know

barge A boat with a flat bottom used for moving goods on a river.

basin A place shaped like a bowl that was cut into the earth by a glacier.

delta A triangle of land made of rich soil left by a river.

evaporate To change from liquid water to a gas.

glacier A large area of thick ice that moves across the land.

reservoir A lake that is made where a dam holds back a river.

••• Contents

Words to Know 2
What Are Lakes and Rivers?.. 4
How Lakes and Rivers Form .. 8
Lakes and Rivers Change
   the Earth 13
Lakes and Rivers Are
   Important 17
Activity: Living Things
   Need Fresh Water 22
Learn More 24
Index 24

What Are Lakes and Rivers?

A lake is a body of water with land all around it. A lake is larger than a pond. Most lakes have fresh water. People need fresh water to live. A river is a large stream of water that flows to the sea. A river may also flow into a lake or into another river. Rivers have fresh water, too.

River Valley

A channel is the path a river takes. Over time, a river channel may cut a wide valley into the land. In a river valley, the river flows at its lowest point. Many people live near lakes and in river valleys.

Lake Louise in Alberta, Canada

A Long Way Down

A waterfall is where a river falls down a high cliff. Angel Falls in Venezuela is the tallest waterfall in the world. It falls 3,212 feet (979 meters).

Saltwater Lakes

Most lakes are fresh water. A small number are salt water, like the oceans. The Great Salt Lake in Utah used to be fresh water. Some of its water **evaporated**. The lake grew more and more salty.

The Dead Sea in Israel and Jordan is not a sea at all. It is a saltwater lake and the world's lowest lake. People like to swim in these lakes. The salty water makes it easy to bob and float in the water.

Angel Falls is located along the Churún River in Venezuela.

How Lakes and Rivers Form

🟢🟣🟡 Lakes and rivers are found in many places. Lakes and rivers may be near an ocean. They may be high in the mountains or down in a valley. They form in different ways.

Lake Huron is the second largest of the Great Lakes. Like all of the Great Lakes, it was formed by melting glaciers.

Water from Glaciers

Most lakes were formed from **glaciers**. The glaciers cut deep **basins** millions of years ago. The basins filled with water when the glaciers melted. The five Great Lakes of North America were made this way.

Melting glaciers made some rivers, too. Glacier water made a small trickle. The water traveled down from high ground. It met other trickles and grew into a stream. Other streams and rivers flowed into it. It became a big river. The Mississippi River was formed this way. It is the longest river in North America. The mighty Mississippi flows 2,340 miles (3,766 kilometers) from north to south. It takes in the Ohio, the Missouri, and the Red Rivers.

The Mississippi River begins at Lake Itasca in Minnesota and flows all the way down to the Gulf of Mexico.

Deep Waters

Lake Baikal is in Russia. It is the deepest lake in the world. More than three hundred rivers flow into Lake Baikal.

Valley Lakes

Earth has a crust covering it. The crust moves. Sometimes the movement of the crust makes a valley. Some places sink. Lakes form in the low places. These valley lakes can be very deep. The Great Rift Valley in Africa has many of them.

Crater Lake in Oregon was formed after a volcano erupted.

A Lake in a Volcano

A crater is like a large bowl where a volcano opens. Lava flows from the opening. When it stops flowing, the crater may fill with water. Crater Lake in Oregon was formed this way. Cameroon, a country in Africa, has thirty-four crater lakes.

Lakes and Rivers Change the Earth

●●● Lakes and rivers change the land. Rivers move water and soil. New lakes form, and others grow bigger or smaller. Water evaporates from lakes and rivers. It falls as rain somewhere on earth.

The Colorado River has worn away the land for millions of years, helping to form the Grand Canyon.

River Canyon

A fast-moving river can cut into the land. It makes a canyon. A canyon is a steep valley shaped like a V. Its walls are rock. The Grand Canyon is in Arizona. It was formed by the Colorado River. It is 1 mile (1.6 km) deep in places. The river flows at the bottom of the canyon. People come from around the world to see the Grand Canyon's beauty.

The Amazon

The Amazon is the largest river in South America. It carries the most water of any river in the world. The Amazon begins in the mountains of Peru. It flows through rain forests. The Amazon carries fresh water to the Atlantic Ocean. It puts 58 billion gallons (219.5 billion liters) of water there every second.

The Amazon River is the second longest in the world, but not a single bridge crosses it. It becomes too wide during the rainy season.

Farmer's Friend

The Nile River flooded every year for thousands of years. The floods brought rich soil. Farmers considered the Nile a power for good.

The Nile

The Nile River is in Africa. It is a huge system of many rivers. The Nile carries soil and sand. The river slows down when it gets to the sea. The river drops its soil and sand. It forms a **delta** in the shape of a fan. The delta soil is rich for planting. Most of Egypt's crops grow there.

Lakes and Rivers Are Important

●●● Thousands of years ago, people began to settle in river valleys. A river gave them water to grow crops. It gave water to raise animals for food. It gave fish to eat. People counted on the river.

River valleys were good places to live. Early communities grew near the Nile River in Egypt. Others grew near the Indus River in India. Some grew near the Tigris and Euphrates Rivers. People built great cities near their rivers. The rivers allowed them to trade with other cities.

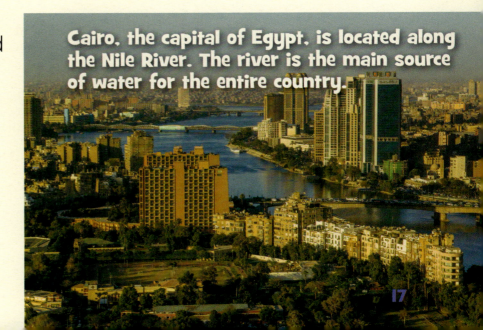

Cairo, the capital of Egypt, is located along the Nile River. The river is the main source of water for the entire country.

India's River

The Ganges is called the mother river of India. Its wide valley is excellent for growing crops. The rice that is grown there feeds hundreds of millions of people.

Dams Make Electricity

For thousands of years, people have built dams. A dam holds back the water in a river. It stops a river from flooding its banks. Hoover Dam in Arizona holds back the Colorado River. The movement of all of the water creates energy. The dam makes electric power for millions of people. It is 726 feet tall (221 m). That is 171 feet (52 m) taller than the Washington Monument in Washington, DC.

Hoover Dam was built in the 1930s. The dam holds back water, which creates Lake Mead (top of photo).

Lakes and rivers provide water, power, and fun activities like rafting.

Reservoir

Lake Mead is a **reservoir** made by Hoover Dam. A reservoir holds a water supply. Lake Mead holds a water supply for millions of people. People in Arizona, California, and Nevada use its water. Lake Mead is the largest reservoir in the United States.

On the Move

Rivers help move goods to the oceans. In the United States, **barges** travel the Mississippi River. They carry loads of crops such as corn and wheat. Then the crops are shipped to places around the world.

Lakes and rivers are important places for many reasons. They supply life-giving water. People enjoy their natural beauty. Clean lakes and rivers are important to everyone. It is up to us to take care of them.

Activity: Living Things Need Fresh Water

Lakes and rivers give us fresh water. Plants, animals, and people are living things. They all need fresh water in order to live.
What would happen if a plant got only salt water?
Let's find out!

- Two glasses half full of water
- Two stalks of celery
- One tablespoon of salt
- Spoon

Step 1: Stir the salt into one glass half full of water. Keep the second glass with only fresh water.

Step 2: Place a stalk of celery in each glass. Leave them for a day.

Step 3: Check the celery stalks.

Questions:

What happened to the celery in salt water?
What happened to the celery in fresh water?
Why?
Salt water took water out of the celery. Fresh water put water in the celery.
If you were a farmer, what would you use to water your crops? Would you use salt water or fresh water?

Plants need fresh water in order to live.

Learn More

Books

Chin, Jason. *Grand Canyon*. New York, NY: Roaring Brook Press, 2017.
Olien, Rebecca. *The Water Cycle at Work*. Mankato, MN: Capstone Press, 2016.
Silverman, Buffy. *Let's Visit the Lake*. Minneapolis, MN: Lerner, 2017.
Spilsbury, Richard. *At Home in Rivers and Lakes*. New York, NY: PowerKids Press, 2016.

Websites

National Geographic Kids: River Otter
kids.nationalgeographic.com/animals/river-otter/#river-otter-swimming.jpg
Information about these playful animals that are found in rivers and lakes.

National Park Service: Grand Canyon National Park
https://www.nps.gov/grca/learn/kidsyouth/index.htm
Explore the park with Sesame Street characters, hear singing park rangers perform, and learn how to become a Junior Park Ranger.

United States Department of Interior: Hoover Dam
www.usbr.gov/lc/hooverdam/educate/kidfacts.html
Amazing facts for kids about Hoover Dam.

Index

Amazon River, 14
basins, 9
canyons, 14
changing the earth, 13–16
cities, 17
crater, 12
crops, 16, 17, 18, 21
dams, 18
delta, 16
formation, 8–12
fresh water, 4, 6, 14, 22
glaciers, 9
Grand Canyon, 14
Mississippi River, 9, 21
Nile River, 16, 17
reservoir, 21
river valley, 4, 17
salt water, 6, 22
valley lakes, 11
volcano, 12

Published in 2020 by Enslow Publishing, LLC.
101 W. 23rd Street, Suite 240, New York, NY 10011

Copyright © 2020 by Enslow Publishing, LLC.

All rights reserved.

No part of this book may be reproduced by any means without the written permission of the publisher.

Library of Congress Cataloging-in-Publication Data
Names: Allman, Barbara, author.
Title: Lakes and rivers / Barbara Allman.
Description: New York, NY : Enslow Publishing, 2020. | Series: Investigate earth science | Includes bibliographical references and index. | Audience:Grades K-4.
Identifiers: LCCN 2018048405| ISBN 9781978507449 (library bound) | ISBN 9781978508682 (pbk.) | ISBN 9781978508699 (6 pack)
Subjects: LCSH: Lakes—Juvenile literature. | Rivers—Juvenile literature.
Classification: LCC GB1603.8 .A44 2020 | DDC 551.48/2—dc23
LC record available at https://lccn.loc.gov/2018048405

Printed in the United States of America

To Our Readers: We have done our best to make sure all website addresses in this book were active and appropriate when we went to press. However, the author and the publisher have no control over and assume no liability for the material available on those websites or on any websites they may link to. Any comments or suggestions can be sent by e-mail to customerservice@enslow.com.

Photos Credits: Cover, p. 1 Jaros/Shutterstock.com; pp. 3, 7 Vadim Petrakov/Shutterstock.com; pp. 3, 8–9 EB Adventure Photography/Shutterstock.com; pp. 3, 12 Stephen Moehle/Shutterstock.com; pp. 3, 15 s.tomas/Shutterstock.com; pp. 3, 17 Prin Adulyatham/Shutterstock.com; p. 5 Olivija Stoilova/EyeEm/Getty Images; p. 10 Phil Schermeister/National Geographic/Getty Images; p. 13 Pete Mcbride/National Geographic/Getty Images; p. 19 scullydion/Shutterstock.com; pp. 20–21 Max Topchii/Shutterstock.com; p. 23 Ulza/Shutterstock.com; cover graphics blackpencil/Shutterstock.com.